Science, Mind & Cosmos

William H. Calvin
Paul Davies
Stephen Jay Gould
W. Daniel Hillis
Steve Jones
Lee Smolin

A Phoenix Paperback

First published in Great Britain by Weidenfeld & Nicolson in 1995
under the title *How Things Are*, edited by
John Brockman and Katinka Matson

This abridged edition published in 1996 by Phoenix
a division of Orion Books Ltd
Orion House, 5 Upper St Martin's Lane, London WC2H 9EA

This abridged edition of *How Things Are* contains the
essays by William H. Calvin, Paul Davies, Stephen Jay Gould,
W. Daniel Hillis, Steve Jones and Lee Smolin.

ISBN 1 85799 585 6

Typeset by CentraCet Ltd, Cambridge
Printed in Great Britain by
Clays Ltd, St Ives plc

Contents

PAUL DAVIES

What Happened Before the Big Bang?

Well what did happen before the big bang?
Few schoolchildren have failed to frustrate their parents with questions of this sort. It often starts with puzzlement over whether space 'goes on forever,' or where humans came from, or how the planet Earth formed. In the end, the line of questioning always seems to get back to the ultimate origin of things: the big bang. 'But what caused *that*?'

Children grow up with an intuitive sense of cause and effect. Events in the physical world aren't supposed to 'just happen.' Something makes them happen. Even when the rabbit appears convincingly from the hat, trickery is suspected. So could the entire universe simply pop into existence, magically, for no actual reason at all?

This simple, schoolchild query has exercised the intellects of generations of philosophers, scientists, and theologians. Many have avoided it as an impenetrable mystery. Others have tried to define it away. Most have got themselves into an awful tangle just thinking about it.

The problem, at rock bottom, is this: If nothing happens without a cause, then *something* must have caused the universe to appear. But then we are faced with the inevitable question of what caused that *something*. And so on in an infinite regress. Some people simply proclaim that God created the universe, but children always want to know who created God, and *that* line of questioning gets uncomfortably difficult.

One evasive tactic is to claim that the universe didn't *have* a beginning, that it has existed for all eternity. Unfortunately, there are many scientific reasons why this obvious idea is unsound. For starters, given an infinite amount of time, anything that can happen will already have happened, for if a physical process is likely to occur with a certain nonzero probability – however small – then given an infinite amount of time the process *must* occur, with probability one. By now, the universe should have reached some sort of final state in which all possible physical processes have run their course. Furthermore, you don't *explain* the existence of the universe by asserting that it has always existed. That is rather like saying that nobody wrote the Bible: it was just copied from earlier versions. Quite apart from all this, there is very good evidence that the universe *did* come into existence in a big bang, about fifteen billion years ago. The effects of that primeval explosion are clearly detectable today – in the fact that the universe is still expanding, and is filled with an afterglow of radiant heat.

So we are faced with the problem of what happened beforehand to trigger the big bang. Journalists love to taunt scientists with this question when they complain about the money being spent on science. Actually, the answer (in my opinion) was spotted a long time ago, by one Augustine of Hippo, a Christian saint who lived in the fifth century. In those days before science, cosmology was a branch of theology, and the taunt came not from journalists, but from pagans: 'What was God doing before he made the universe?' they asked. 'Busy creating Hell for the likes of you!' was the standard reply.

But Augustine was more subtle. The world, he claimed, was made 'not *in* time, but simultaneously *with* time.'

In other words, the origin of the universe – what we now call the big bang – was not simply the sudden appearance of matter in an eternally preexisting void, but the coming into being of time itself. Time *began* with the cosmic origin. There was no 'before,' no endless ocean of rime for a god, or a physical process, to wear itself out in infinite preparation.

Remarkably, modern science has arrived at more or less the same conclusion as Augustine, based on what we now know about the nature of space, time, and gravitation. It was Albert Einstein who taught us that time and space are not merely an immutable arena in which the great cosmic drama is acted out, but are part of the cast – part of the physical universe. As physical entities, time and space can change – suffer distortions – as a result of gravitational processes. Gravitational theory predicts that under the extreme conditions that prevailed in the early universe, space and time may have been so distorted that there existed a boundary, or 'singularity,' at which the distortion of space-time was infinite, and therefore through which space and time cannot have continued. Thus, physics predicts that time was indeed bounded in the past as Augustine claimed. It did not stretch back for all eternity.

If the big bang was the beginning of time itself, then any discussion about what happened before the big bang, or what caused it – in the usual sense of physical causation – is simply meaningless. Unfortunately, many children, and adults, too, regard this answer as disingenuous. There must be more to it than that, they object.

Indeed there is. After all, *why* should time suddenly 'switch on'? What explanation can be given for such a singular event? Until recently, it seemed that any explanation of the initial 'singularity' that marked the origin of time would have to lie beyond the scope of science. However, it all depends on what is meant by 'explanation.' As I remarked, all children have a good idea of the notion of cause and effect, and usually an explanation of an event entails finding something that caused it. It turns out, however, that there are physical events which do *not* have well-defined causes in the manner of the everyday world. These events belong to a weird branch of scientific inquiry called quantum physics.

Mostly, quantum events occur at the atomic level; we don't experience them in daily life. On the scale of atoms and molecules, the usual commonsense rules of cause and effect are suspended. The rule of law is replaced by a sort of anarchy or chaos, and things happen *spontaneously* – for no particular reason. Particles of matter may simply pop into existence without warning, and then equally abruptly disappear again. Or a particle in one place may suddenly materialize in another place, or reverse its direction of motion. Again, these are real effects occurring on an atomic scale, and they can be demonstrated experimentally.

A typical quantum process is the decay of a radioactive nucleus. If you ask why a given nucleus decayed at one particular moment rather than at some other, there is no answer. The event 'just happened' at that moment, that's all. You cannot predict these occurrences. All you can do is give the probability – there is a fifty-fifty chance that a given nucleus will decay in, say, one hour. This uncertainty is not simply a result of our ignorance of all the little forces and influences that try to make the nucleus decay; it is inherent in nature itself, a basic part of quantum reality.

The lesson of quantum physics is this: Something that 'just happens' need not actually violate the laws of physics. The abrupt and uncaused appearance of something can occur within the scope of scientific law, once quantum laws have been taken into account. Nature apparently has the capacity for genuine spontaneity.

It is, of course, a big step from the spontaneous and uncaused appearance of a subatomic particle – something that is routinely observed in particle accelerators – to the spontaneous and uncaused appearance of the universe. But the loophole is there. If, as astronomers believe, the primeval universe was compressed to a very small size, then quantum effects must have once been important on a cosmic scale. Even if we don't have a precise idea of exactly what took place at the beginning, we can at least see that the origin of the universe from nothing need not be unlawful or unnatural or unscientific. In short, it need not have been a supernatural event.

Inevitably, scientists will not be content to leave it at that. We would like to flesh out the details of this profound concept. There is even a subject devoted to it, called quantum cosmology. Two famous quantum cosmologists, James Hartle and Stephen Hawking, came up with a clever idea that goes back to Einstein. Einstein not only found that space and time are part of the physical universe; he also found that they are linked in a very intimate way. In fact, space on its own and time on its own are no longer properly valid concepts. Instead, we must deal with a unified 'space-time' continuum. Space has three dimensions, and time has one, so space-time is a four-dimensional continuum.

In spite of the space-time linkage, however, space is space and time is time under almost all circumstances. Whatever space-time distortions gravitation may produce, they never turn space into time or time into space. An exception arises, though, when quantum effects are taken into account. That all-important intrinsic uncertainty that afflicts quantum systems can be applied to space-time, too. In this case, the uncertainty can, under special circumstances, affect the *identities* of space and time. For a very, very brief duration, it is possible for time and space to merge in identity, for time to become, so to speak, spacelike – just another dimension of space.

The spatialization of time is not something abrupt; it is a continuous process. Viewed in reverse as the temporalization of (one dimension of) space, it implies that time can *emerge out of space* in a continuous process. (By continuous, I mean that the timelike quality of a dimension, as opposed to its spacelike quality, is not an all-or-nothing affair; there are shades in between. This vague statement can be made quite precise mathematically.)

The essence of the Hartle-Hawking idea is that the big bang was not the abrupt switching on of time at some singular first moment, but the emergence of time from space in an ultrarapid but nevertheless continuous manner. On a human time scale, the big bang was very much a sudden, explosive origin of space, time, and matter. But look very, very closely at that first tiny fraction of a second and you find that there was no precise and sudden beginning at all. So here we have a theory of the origin of the universe that seems to say two contradictory things: First, time did not always exist; and 7 second, there was no first moment of time. Such are the oddities of quantum physics.

Even with these further details thrown in, many people feel cheated. They want to ask *why* these weird things happened, *why* there is a universe, and why *this* universe. Perhaps science cannot answer such questions. Science is good at telling us how, but not so good on the why. Maybe there isn't a why. To wonder why is very human, but perhaps there is no answer in human terms to such deep questions of existence. Or perhaps there is, but we are looking at the problem in the wrong way.

Well, I didn't promise to provide the answers to life, the universe, and everything, but I have at least given a plausible answer to the question I started out with: What happened before the big bang?

The answer is: Nothing.

STEPHEN JAY GOULD

Three Facets of Evolution

1. WHAT EVOLUTION IS NOT

Of all the fundamental concepts in the life sciences, evolution is both the most important and the most widely misunderstood. Since we often grasp a subject best by recognising what it isn't, and what it cannot do, we should begin with some disclaimers, acknowledging for science what G. K. Chesterton considered so important for the humanities: 'Art is limitation; the essence of every picture is the frame.'

First, neither evolution, nor any science, can access the subject of ultimate origins or ethical meanings. (Science, as an enterprise, tries to discover and explain the phenomena and regularities of the empirical world, under the assumption that natural laws are uniform in space and time. This restriction places an endless world of fascination within the 'picture'; most subjects thus relegated to the 'frame' are unanswerable in any case.) Thus, evolution is not the study of life's ultimate origin in the universe or of life's intrinsic significance among nature's objects; these questions are philosophical (or theological) and do not fall within the purview of science. (I also suspect that they have no universally satisfactory answers, but this is another subject for another time.) This point is important because zealous fundamentalists, masquerading as 'scientific creationists,' claim that creation must be equated with evolution, and be given equal time in schools, because both are equally 'religious' in dealing with ultimate unknowns. In fact, evolution does not treat such subjects at all, and thus remains fully scientific.

Second, evolution has been saddled with a suite of concepts and meanings that represent long-standing Western social prejudices and psychological hopes, rather than any account of nature's factuality. Such 'baggage' may be unavoidable for any field so closely allied with such deep human concerns (see part 3 of this statement), but this strong social overlay has prevented us from truly completing Darwin's revolution. Most pernicious and constraining among these prejudices is the concept of progress, the idea that evolution possesses a driving force or manifests an overarching trend toward increasing complexity, better biomechanical design, bigger brains, or some other parochial definition of progress centered upon a long-standing human desire to place ourselves atop nature's pile – and thereby assert a natural right to rule and exploit our planet.

10   Evolution, in Darwin's formulation, is adaptation to

changing local environments, not universal 'progress.' A lineage of elephants that evolves a heavier coating of hair to become a woolly mammoth as the ice sheets advance does not become a superior elephant in any general sense, but just an elephant better adapted to local conditions of increasing cold. For every species that does become more complex as an adaptation to its own environment, look for parasites (often several species) living within its body – for parasites are usually greatly simplified in anatomy compared with their freeliving ancestors, yet these parasites are as well adapted to the internal environment of their host as the host has evolved to match the needs of its external environment.

2. WHAT EVOLUTION IS

In its minimalist, 'bare bones' formulation, evolution is a simple idea with a remarkable range of implications. The basic claim includes two linked statements that provide rationales for the two central disciplines of natural history: taxonomy (or the order of relationships among organisms), and paleontology (or the history of life). Evolution means (1) that all organisms are related by ties of genealogy or descent from common ancestry along the branching patterns of life's tree, and (2) that lineages alter their form and diversity through time by a natural process of change – 'descent with modification' in 11

Darwin's chosen phrase. This simple, yet profound, insight immediately answers the great biological question of the ages: What is the basis for the 'natural system' of relationships among organisms (cats closer to dogs than to lizards; all vertebrates closer to each other than any to an insect – a fact well appreciated, and regarded as both wonderful and mysterious, long before evolution provided the reason). Previous explanations were unsatisfactory because they were either untestable (God's creative hand making each species by fiat, with taxonomic relationships representing the order of divine thought), or arcane and complex (species as natural places, like chemical elements in the periodic table, for the arrangement of organic matter). Evolution's explanation for the natural system is so stunningly simple: Relationship is genealogy; humans are like apes because we share such a recent common ancestor. The taxonomic order is a record of history.

But the basic fact of genealogy and change – descent with modification – is not enough to characterise evolution as a science. For science has two missions: (1) to record and discover the factual state of the empirical world, and (2) to devise and test explanations for why the world works as it does. Genealogy and change only represent the solution to this first goal – a description of the fact of evolution. We also need to know the mechanisms by which evolutionary change occurs – the second

goal of explaining the causes of descent with modification. Darwin proposed the most famous and best-documented mechanism for change in the principle that he named 'natural selection.'

The fact of evolution is as well documented as anything we know in science – as secure as our conviction that Earth revolves about the sun, and not vice versa. The mechanism of evolution remains a subject of exciting controversy – and science is most lively and fruitful when engaged in fundamental debates about the causes of well-documented facts. Darwin's natural selection has been affirmed, in studies both copious and elegant, as a powerful mechanism, particularly in evolving the adaptations of organisms to their local environments – what Darwin called 'that perfection of structure and coadaptation which most justly excites our admiration.' But the broad-scale history of life includes other phenomena that may require different kinds of causes as well (potentially random effects, for example, in another fundamental determinant of life's pattern – which groups live, and which die, in episodes of catastrophic extinction).

3. WHY SHOULD WE CARE?

The deepest, in-the-gut, answer to this question lies in the human psyche, and for reasons that I cannot begin to fathom. We are fascinated by physical ties of ancestry; we feel that we will understand ourselves better, know who we are in some fundamental sense, when we trace the sources of our descent. We haunt graveyards and parish records; we pore over family Bibles and search out elderly relatives, all to fill in the blanks on our family tree. Evolution is this same phenomenon on a much more inclusive scale – roots writ large. Evolution is the family tree of our races, species, and lineages – not just of our little, local surname. Evolution answers, insofar as science can address such questions at all, the troubling and fascinating issues of 'Who are we?' 'To which other creatures are we related, and how?' 'What is the history of our interdependency with the natural world?' 'Why are we here at all?' Beyond this, I think that the importance of evolution in human thought is best captured in a famous statement by Sigmund Freud, who observed, with wry and telling irony, that all great scientific revolutions have but one feature in common: the casting of human arrogance off one pedestal after another of previous convictions about our ruling centrality in the universe. Freud mentions three such revolutions: the Copernican, for moving our home from center stage in a small universe to a tiny peripheral hunk of rock amid inconceivable vastness; the Darwinian, for 'relegating us to descent from an animal world'; and (in one of the least modest statements of intellectual history) his own, for discovering the unconscious and illustrating the nonrationality of the human mind. What can be more humbling, and therefore more liberating, than a transition from viewing ourselves as 'just a little lower than the angels,' the created rulers of nature, made in God's image to shape and subdue the earth – to the knowledge that we are not only natural products of a universal process of descent with modification (and thus kin to all other creatures), but also a small, late-blooming, and ultimately transient twig on the copiously arborescent tree of life, and not the foreordained summit of a ladder of progress. Shake complacent certainty, and kindle the fires of intellect.

Why Are Some People Black?

Everyone knows – do they not? – that many people have black skin. What is more, black people are concentrated in certain places – most notably, in Africa – and, until the upheavals of the past few centuries, they were rare in Europe, Asia, and the Americas. Why should this be so?

It seems a simple question. Surely, if we cannot give a simple answer, there is something wrong with our understanding of ourselves. In fact, there is no straightforward explanation of this striking fact about humankind. Its absence says a lot about the strengths and weaknesses of the theory of evolution and of what science can and cannot say about the past.

Any anatomy book gives one explanation of why people look different. Doctors love pompous words, particularly if they refer to other doctors who lived long ago. Black people have black skin, their textbooks say, because they have a distinctive Malpighian layer. This is a section of the skin named after the seventeenth-century Italian anatomist Malpighii. It contains lots of cells called *melanocytes*. Within them is a dark pigment called melanin. The more there is, the blacker the skin. Malpighii found that African skin had more melanin than did that of Europeans. The question was, it seemed, solved.

This is an example of what I sometimes think of as 'the Piccadilly explanation.' One of the main roads in London is called Piccadilly – an oddly un-English word. I have an amusing book that explains how London's streets got their names. What it says about Piccadilly sums up the weakness of explanations that depend, like the anatomists', only on describing a problem in more detail. The street is named, it says, after the tailors who once lived there and made high collars called *piccadills*. Well, fair enough; but surely that leaves the interesting question unanswered. Why call a collar a piccadill in the first place? It is not an obvious word for an everyday piece of clothing. My book is, alas, silent.

Malphighii's explanation may be good enough for doctors, but will not satisfy any thinking person. It answers the question *how* but not the more interesting question *why* there is more melanin in African skin.

Because the parents, grandparents, and – presumably – distant ancestors of black people are black, and those of white people white, the solution must lie in the past. And that is a difficulty for the scientific method. It is impossible to check directly just what was going on when the first blacks or the first whites appeared on earth. Instead, we must depend on indirect evidence.

There is one theory that is, if nothing else, simple and consistent. It has been arrived at again and again. It depends solely on belief; and if there is belief, the question of proof does not arise. Because of this, the theory lies outside science.

It is that each group was separately created by divine action. The Judeo-Christian version has it that Adam and Eve were created in the Garden of Eden. Later, there was a gigantic flood; only one couple, the Noahs, survived. They had children: Ham, Shem, and Japhet. Each gave rise to a distinct branch of the human race, Shem to the Semites, for example. The children of Ham had dark skins. From them sprang the peoples of Africa. That, to many people, is enough to answer the question posed in this essay.

The Noah story is just a bald statement about history. Some creation myths are closer to science. They try to *explain* why people look different. One African version is that God formed men from clay, breathing life into his creation after it had been baked. Only the Africans were fully cooked – they were black. Europeans were not quite finished and were an unsatisfactory muddy pink.

The trouble with such ideas is that they cannot be disproved. I get lots of letters from people who believe passionately that life, in all its diversity, appeared on earth just a few thousand years ago as a direct result of God's intervention. There is no testimony that can per-suade them otherwise. Prove that there were dinosaurs millions of years before humans, and they come up with rock 'footprints' showing, they say, that men and dinosaurs lived together as friends. So convinced are they of the truth that they insist that their views appear in school textbooks.

If all evidence, whatever it is, can only be interpreted as supporting one theory, then there is no point in arguing. In fact, if belief in the theory is strong enough, there is no point in looking for evidence in the first place. Certainty is what blocked science for centuries. Scientists are, if nothing else, uncertain. Their ideas must constantly be tested against new knowledge. If they fail the test, they are rejected.

No biologist now believes that humans were created through some miraculous act. All are convinced that they evolved from earlier forms of life. Although the proof of the fact of evolution is overwhelming, there is plenty of room for controversy about how it happened. Nowhere is this clearer than in the debate about skin color.

Modern evolutionary biology began with the nineteenth-century English biologist Charles Darwin. He formed his ideas after studying geology. In his day, many people assumed that grand features such as mountain ranges or deep valleys could arise only through sudden catastrophes such as earthquakes or volcanic eruptions, which were unlikely to be seen by scientists as they were so rare. Darwin realized that, given enough time, even a small stream can, by gradually wearing away the rocks, carve a deep canyon. The present, he said, is the key to the past. By looking at what is going on in a landscape today, it is possible to infer the events of millions of years ago. In the same way, the study of living creatures can show what happened in evolution.

In *The Origin of Species*, published in 1859, Darwin suggested a mechanism whereby new forms of life could evolve. *Descent with modification*, as he called it, is a simple piece of machinery, with two main parts.

One produces inherited diversity. This process is now known as mutation. In each generation, there is a small but noticeable chance of a mistake in copying genes as sperm or egg are made. Sometimes we can see the results of mutations in skin color; one person in several thousand is an albino, lacking all skin pigment. Albinos are found all over the world, including Africa. They descend from sperm or eggs that have suffered damage in the pigment genes.

The second piece of the machine is a filter. It separates mutations which are good at coping with what the environment throws at them from those which are not. Most mutations – albinism, for example – are harmful. The people who carry mutant genes are less likely to survive and to have children than do those who do not. Such mutations quickly disappear. Sometimes, though, one turns up which is better at handling life's hardships than what went before. Perhaps the environment is changing, or perhaps the altered gene simply does its job better. Those who inherit it are more likely to survive; they have more children, and the gene becomes more common. By this simple mechanism, the population has evolved through *natural selection*. Evolution, thought Darwin, was a series of successful mistakes.

If Darwin's machine worked for long enough, then new forms of life – new species – would appear. Given enough time, all life's diversity could emerge from simple ancestors. There was no need to conjure up ancient and unique events (such as a single incident of creation) which could neither be studied nor duplicated. Instead, the living world was itself evidence for the workings of evolution.

What does Darwin's machine tell us about skin color? As so often in biology, what we have is a series of intriguing clues, rather than a complete explanation.

There are several kinds of evidence about how things evolve. The best is from fossils: the preserved remnants of ancient times. These contain within themselves a statement of their age. The chemical composition of bones (or of the rocks into which they are transformed) shifts with time. The molecules decay at a known rate, and certain radioactive substances change from one form into another. This gives a clue as to when the original owner of the bones died. It may be possible to trace the history of a family of extinct creatures in the changes that occur as new fossils succeed old.

The human fossil record is not good – much worse, for

example, than that of horses. In spite of some enormous gaps, enough survives to make it clear that creatures looking not too different from ourselves first appeared around a hundred and fifty thousand years ago. Long before that, there were apelike animals which looked noticeably human but would not be accepted as belonging to our own species if they were alive today. No one has traced an uninterrupted connection between these extinct animals and ourselves. Nevertheless, the evidence for ancient creatures that changed into modern humans is overwhelming.

As there are no fossilized human skins, fossils say nothing directly about skin color. They do show that the first modern humans appeared in Africa. Modern Africans are black. Perhaps, then, black skin evolved before white. Those parts of the world in which people have light skins – northern Europe, for example – were not populated until about a hundred thousand years ago, so that white skin evolved quite quickly.

Darwin suggested another way of inferring what happened in the past: to compare creatures living today. If two species share a similar anatomy, they probably split from their common ancestor more recently than did another which has a different body plan. Sometimes it is possible to guess at the structure of an extinct creature by looking at its living descendants.

This approach can be used not just for bones but for
molecules such as DNA. Many biologists believe that

DNA evolves at a regular rate: that in each generation, a small but predictable proportion of its subunits changes from one form into another. If this is true (and often it is), then counting the changes between two species reveals how closely they are related. What is more, if they share an ancestor that has been dated using fossils, it allows DNA to be used as a 'molecular clock,' timing the speed of evolution. The rate at which the clock ticks can then be used to work out when other species split by comparing their DNA, even if no fossils are available.

Chimpanzees and gorillas seem, from their body plan, to be our relatives. Their genes suggest the same thing. In fact, each shares 98 percent of its DNA with ourselves, showing just how recently we separated. The clock suggests that the split was about six million years ago. Both chimp and gorilla have black skins. This, too, suggests that the first humans were black and that white skin evolved later.

However, it does not explain *why* white skin evolved. The only hint from fossils and chimps is that the change took place when humans moved away from the tropics. We are, without doubt, basically tropical animals. It is much harder for men and women to deal with cold than with heat. Perhaps climate has something to do with skin color.

To check this idea, we must, like Darwin, look at living creatures. Why should black skin be favored in hot and sunny places and white where it is cool and cloudy? It is easy to come up with theories, some of which sound pretty convincing. However, it is much harder to test them.

The most obvious idea is wrong. It is that black skin protects against heat. Anyone who sits on a black iron bench on a hot sunny day soon discovers that black objects heat up *more* than white ones do when exposed to the sun. This is because they absorb more solar energy. The sun rules the lives of many creatures. Lizards shuttle back and forth between sun and shade. In the California desert, if they stray more than six feet from shelter on a hot day, they die of heat stroke before they can get back. African savannahs are dead places at noon, when most animals are hiding in the shade because they cannot cope with the sun. In many creatures, populations from hot places are lighter – not darker – in color to reduce the absorption of solar energy.

People, too, find it hard to tolerate full sunshine – blacks more so than whites. Black skin does not protect those who bear it from the sun's heat. Instead, it makes the problem worse.

However, with a bit of ingenuity, it is possible to bend the theory slightly to make it fit. Perhaps it pays to have black skin in the chill of the African dawn, when people begin to warm up after a night's sleep. In the blaze of noon, one can always find shelter under a tree.

The sun's rays are powerful things. They damage the skin. Melanin helps to combat this. One of the first signs of injury is an unhealthy tan. The skin is laying down an emergency layer of melanin pigment. Those with fair skin are at much greater risk from skin cancer than are those with dark. The disease reaches its peak in Queensland, in Australia, where fair-skinned people expose themselves to a powerful sun by lying on the beach.

Surely, this is why black skin is common in sunny places – but, once again, a little thought shows that it probably is not. Malignant melanoma, the most dangerous skin cancer, may be a vicious disease, but it is an affliction of middle age. It kills its victims after they have passed on their skin-color genes to their children. Natural selection is much more effective if it kills early in life. If children fail the survival test, then their genes perish with their carriers. The death of an old person is irrelevant, as their genes (for skin color or anything else) have already been handed on to the next generation.

The skin is an organ in its own right, doing many surprising things. One is to synthesize vitamin D. Without this, children suffer from rickets: soft, flexible bones. We get most vitamins (essential chemicals needed in minute amounts) from food. Vitamin D is unusual. It can be made in the skin by the action of sunlight on a natural body chemical. To do this, the sun must get into the body. Black people in sunshine hence make much less vitamin D than do those with fair skins. Vitamin D is particularly important for children, which is why babies (African or European) are lighter in color than are adults. 25

Presumably, then, genes for relatively light skin were favored during the spread from Africa into the cloud and rain of the north. That might explain why Europeans are white – but does it reveal why Africans are black? Too much vitamin D is dangerous (as some people who take vitamin pills discover to their cost). However, even the fairest skin cannot make enough to cause harm. The role of black skin is not to protect against excess Vitamin D.

It may, though, be important in preserving other vitamins. The blood travels around the body every few minutes. On the way, it passes near the surface of the skin through fine blood vessels. There, it is exposed to the damaging effects of the sun. The rays destroy vitamins – so much so, that a keen blond sunbather is in danger of vitamin deficiency. Even worse, the penetrating sunlight damages antibodies, the defensive proteins made by the immune system. In Africa, where infections are common and, sometimes, food is short, vitamin balance and the immune system are already under strain. The burden imposed by penetrating sunlight may be enough to tip the balance between health and disease. Dark skin pigment may be essential for survival. No one has yet shown directly whether this is true.

There are plenty of other theories as to why some people are black. For an African escaping from the sun under a tree, black skin is a perfect camouflage. Sexual preference might even have something to do with the evolution of skin color. If, for one reason or another, people choose their partners on the basis of color, then the most attractive genes will be passed on more effectively. A slight (and perhaps quite accidental) preference for dark skin in Africa and light in Europe would be enough to do the job. This kind of thing certainly goes on with peacocks – in which females prefer males with brightly patterned tails – but there is no evidence that it happens in humans.

Accident might be important in another way, too. Probably only a few people escaped from Africa a hundred thousand years and more ago. If, by chance, some of them carried genes for relatively light skins, then part of the difference in appearance between Africans and their northern descendants results from a simple fluke. There is a village of North American Indians today where albinos are common. By chance, one of the small number of people who founded the community long ago carried the albino mutation and it is still abundant there.

All this apparent confusion shows how difficult it is for science to reconstruct history. Science is supposed to be about testing, and perhaps disproving, hypotheses. As we have seen, there is no shortage of ideas about why people differ in skin color. Perhaps none of the theories is correct; or perhaps one, two, or all of them are. Because whatever gave rise to the differences in skin color in different parts of the world happened long ago, no one can check directly.

But science does not always need direct experimental

tests. A series of indirect clues may be almost as good. The hints that humans evolved from simpler predecessors and are related to other creatures alive today are so persuasive that it is impossible to ignore them. So far, we have too few facts and too many opinions to be certain of all the details of our own evolutionary past. However, the history of the study of evolution makes me confident that, some day, the series of hints outlined in this essay will suddenly turn into a convincing proof of just why some people are black and some white.

William H. Calvin

How to Think What No One Has Ever Thought Before

The short answer is to take a nap and dream about something. Our dreams are full of originality. Their elements are all old things, our memories of the past, but the combinations are original. Combinations make up in variety what they lack in quality, as when we dream about Socrates driving a bus in Brooklyn and talking to Joan of Arc about baseball. Our dreams get time, place, and people all mixed up.

Awake, we have a stream of consciousness, also containing a lot of mistakes. But we can quickly correct those mistakes, usually before speaking out loud. We can improve the sentence, even as we are speaking it. Indeed, most of the sentences we speak are ones we've never spoken before. We construct them on the spot. But *how* do we do it, when we say something we've never said before – and it doesn't come out as garbled as our dreams?

We also forecast the future in a way that no other animal can do. Since it hasn't happened yet, we have to imagine what might happen. We often preempt the future by taking actions to head off what will otherwise happen. We can think before acting, guessing how objects or people might react to a proposed course of action.

That is extraordinary when compared to all other animals. It even takes time to develop in children. By the time children go to school, adults start expecting them to be responsible for predicting the consequences: 'You should have realized that . . .' and 'Think before you do something like that!' aren't seriously said to babies and most preschoolers – or our pets. We don't seriously expect our dogs and cats to appraise a novel situation, like a fish falling out of the refrigerator, with an eye toward saving it for the dinner guests tonight.

An ability to guess the consequences of a course of action is the foundation of ethics. Free will implies not only the choice between known alternatives, but an ability to imagine novel alternatives and to shape them up into something of quality. Many animals use trial and error, but we humans do a great deal of it 'off line,' before actually acting in the real world. The process of contemplation and mental rehearsal that shapes up novel variations would appear to lie at the core of some of our most cherished human attributes. *How* do we do that?

Creating novelty isn't difficult. New arrangements of old things will do.

Everyone thinks that mutations (as when a cosmic ray

comes along and knocks a DNA base out of position, allowing another to fill in) are where novel genes come from. Nature actually has two other mechanisms that are more important: copying errors and shuffling the deck. Anyone with a disk drive knows that copying errors is the way things are, that procedures had to be invented to detect them (such as those pesky check sums) and correct them (such as the error-correcting codes which are now commonplace). All that's required to achieve novelty is to relax vigilance.

But nature occasionally works hard at mixing up things, each time that a sperm or ovum is made; the genes on both chromosomes of a pair are shuffled (what's known as crossing over during meiosis) before being segregated into the new chromosome arrangement of the sperm or ovum. And, of course, fertilization of an ovum by another individual's sperm creates a new third individual, one that has a choice (in most cases) between using a gene inherited from the mother or the version of it inherited from the father.

Quality is the big problem, not novelty as such. Nature's usual approach to quality is to try lots of things and see what works, letting the others fall by the wayside. For example, lots of sperm are defective, missing essential chromosomes; should they fertilize an ovum, development will fail at some point, usually so early that pregnancy isn't noticed. For this and other reasons, over 80 percent of human conceptions fail, most in the first six weeks (this spontaneous abortion rate is far more significant than even the highest rates of induced abortions).

There are often high rates of infant and juvenile mortality as well; only a few individuals of any species survive long enough to become sexually mature and themselves become parents. As Charles Darwin first realized in about 1838, this is a way that plants and animals change over many generations into versions that are better suited to environmental circumstances. Nature throws up a lot of variations with each new generation, and some are better suited to the environment than others. Eventually, a form of quality emerges through this shaping-up process.

When Darwin explained how evolution might produce more and more complex animals, it started the psychologists thinking about thought itself. Might the mind work the same way as Darwin's mechanism for shaping a new species? Might a new thought be shaped by a similar process of variation and selection?

Most random variations on a standard behavior, even if only changing the order of actions, are less efficient, and some are dangerous ('look after you leap'). Again, quality is the problem, not novelty per se. Most animals confine themselves to well-tested solutions inherited from ancestors that survived long enough to successfully reproduce. New combinations are sometimes tried out in play as juveniles, but adults are far less playful.

By 1880, in an article in the *Atlantic Monthly*, the

pioneering American psychologist William James (who invented the literary term 'stream of consciousness') had the basic idea:

> [T]he new conceptions, emotions, and active tendencies which evolve are originally *produced* in the shape of random images, fancies, accidental outbursts of spontaneous variations in the functional activity of the excessively unstable human brain, which the outer environment simply confirms or refutes, preserves or destroys – selects, in short, just as it selects morphological and social variations due to molecular accidents of an analogous sort.

His French contemporary, Paul Souriau, writing in 1881, said much the same thing:

> We know how the series of our thoughts must end, but ... it is evident that there is no way to begin except at random. Our mind takes up the first path that it finds open before it, perceives that it is a false route, retraces its steps and takes another direction. ... By a kind of artificial selection, we can ... substantially perfect our own thought and make it more and more logical.

James and Souriau were building on the even more basic idea of Alexander Bain, concerning trial and error. Writing in Scotland in 1855, Bain initially employed the phrase *trial and error* when considering the mastery of 33 motor skills such as swimming. Through persistent effort, Bain said, the swimmer stumbles upon the 'happy combination' of required movements and can then proceed to practice them. He suggested that the swimmer needed a sense of the effect to be produced, a command of the elements, and that he then used trial and error until the desired effect was actually produced. This is what a Darwinian process can use to shape up a thought – which is, after all, a plan for a movement, such as what to say next.

Surprisingly, no one seemed to know what to do next, to make the link between the basic idea of Darwinian thought and the rest of psychology and neurobiology. For more than a century, this key idea has lain around like a seed in poor soil, trying to take hold. One problem is that it is easy (even for scientists) to adopt a cartoon version of Darwinism – survival of the fittest, or selective survival – and fail to appreciate the rest of the process.

The basic Darwinian idea is deceptively simple. Animals always reproduce, but all of their offspring don't manage to grow up to have babies themselves – they overproduce. There is a lot of variation in the offspring of the same two parents; each offspring (identical twins and clones excepted) gets a different set of shuffled chromosomes.

Operating on this generated diversity is selective survival. Some variants survive into adulthood better than others, and so, the next generation's variations are based on the survivor's genes. Some are better, most are worse, but they center around an advanced position because the worst ones tend to die young. And the next generation is, for the average survivor into adulthood, even better suited to the environment's particular collection of food, climate, predators, nesting sites, etc.

We usually think in terms of millennia as the time scale of this process that can evolve a new species. With artificial selection by animal breeders, substantial effects can be produced in a dozen generations. But the process can operate on the time scale of the immune response, as new antibodies are shaped up by success in killing off invading molecules; within a week or two, antibody shapes can be evolved that have a key-and-lock specificity for a foreign molecule. Might the same process suffice for the time scale of thought and action?

It is worth restating the six essentials of a Darwinian process a bit more abstractly, so we can separate the principles from the particulars:

- There is a pattern involved (typically, a string of DNA bases – but the pattern could also be a musical melody or the brain pattern associated with a thought).
- Copies are somehow made of this pattern (as when cells divide, but also when someone whistles a tune he's heard).
- Variations on the pattern sometimes occur, whether by copying errors or by shuffling the deck.

35

- Variant patterns compete for occupation of a limited space (as when bluegrass and crabgrass compete for your backyard).

- The relative success of the variant patterns is influenced by a multifaceted environment (for grass, it's hours of sunlight, soil nutrients, how often it's watered, how often it's cut, etc.).

- And, most important, the process has a loop. The next generation is based on which variants survived to maturity, and that shifts the base from which the surviving variants spread their own reproductive bets. And the next, and the next. This differential survival means that the variation process is not truly random. Instead, it is based on those patterns that have survived the multifaceted environment's selection process. A spread around the currently successful is created; most will be worse, but some may be better.

Not every process that makes copies of patterns is going to qualify as Darwinian. Photocopy and fax machines make copies of the ink patterns on a sheet of paper, but there is usually no loop.

If you do make copies of copies, for dozens of generations, you will see some copying errors (especially if copying grayscale photographs). Now you've satisfied the first three conditions – but you still don't have competition for a work space that is biased by a multifaceted environment, nor an advantage in reproduction for certain variants.

Similarly, you can have selective survival without the rest of the Darwinian process. You will find more fifteen-year-old Volvos still on the road than you will Fiats of the same age. But the fifteen-year-old Volvo doesn't reproduce. Nor do brain cells – though the connections between them (synapses) are edited over time. In the brain of an infant, there are many connections between nerve cells that don't survive into adulthood. But this selective survival (random connections that prove to be useful) isn't proper Darwinism either, unless the surviving connection patterns somehow manage to reproduce themselves elsewhere in that brain (or perhaps through mimicry in someone else's brain). And, even if they did, this pattern copying would still have to satisfy the requirement for a loop where reproduction with new variation is biased toward the more successful.

Selective survival is a powerful mechanism that produces crystals in nonliving matter as well as economic patterns in cultural evolution. Selective survival is a problem for all business enterprises, especially small ones, but it leads – at least in capitalist free-market theories – to a better fit with 'what works.'

Selective survival of all sorts is sometimes called Darwinian (Darwin was annoyed when Herbert Spencer started talking of social Darwinism). But selective survival per se can even be seen in nonliving systems, as when flowing water carries away the sand grains and leaves the pebbles behind on a beach.

Full-fledged Darwinism is even more powerful, but it requires differential reproduction of the more successful. Economics has some recent examples in fast-food franchises, where copies are produced of the more successful of an earlier generation. Indeed, they seem to be in a competition with their variants for a limited 'work space.' If they close the loop by generating new variations on the more successful (imagine a MacUpscale and a Mac-Economy splitting off from one of the chains), they may provide another example of a Darwinian process evolving new complexity.

When people call something 'Darwinian,' they're usually referring to only part of the Darwinian process, something that uses only several of the six essentials. And, so powerful are the words we use, this overly loose terminology has meant that people (scientists included) haven't realized what was left out.

Indeed, the second reason why the Darwinian thought idea wasn't fleshed out earlier is that it has taken a while to realize that thought patterns might need to be copied – and that the copies might need to compete with copies of alternative thoughts. Since we haven't known how to describe the neural activities underlying thought, we haven't been able to think about copying. But copying is 38 a major clue about what the thought process must be like; it's a constraint that considerably reduces the possibilities.

In the early 1950s, during the search for the genetic code, molecular biologists were acutely aware of the need for a molecular process that could somehow make copies of itself during cell division. The reason why the double helix structure was so satisfying in 1953 was that it solved the copying problem. In subsequent years, the genetic code (the translation table between DNA triplets and amino acid strings) was worked out. Perhaps we too can identify the cerebral code that represents an object or idea, with the aid of looking at what cerebral patterns can be semiaccurately copied.

Thoughts are just combinations of sensations and memories – or, looked at another way, thoughts are movements that haven't happened yet (and maybe never will). The brain produces movements with a barrage of nerve impulses going to the muscles, whether limbs or larynx. But what determines the details of this barrage?

Sometimes, it is simply an innate rhythm such as the ones which produce chewing, breathing, and walking. Sometimes there is time for lots of corrections, as when you lift a coffee cup and discover that it weighs less than you thought; before it hits your nose, you manage to make some corrections to your arm muscles. But some movements are so quick (over and done in one eighth of a second) that no feedback is possible: throwing, hammering, clubbing, kicking, spitting (including 'spitting out 39 a word'). We call these ballistic movements; they're particularly interesting because they require that a complete plan be evolved before acting. During 'get set,' you have to produce the perfect plan. A plan for a movement is like the roll for a player piano: eighty-eight output channels, one for each key, and the times at which each key is struck. To hammer or throw indeed requires coordinating close to eighty-eight muscles, so think of a sheet of music as a plan for a spatiotemporal pattern – all of those chords, melodies, and interweaving patterns we call musical.

In 1949, the Canadian psychologist Donald Hebb formulated his cell-assembly hypothesis, stating that evoking a memory required reconstituting a pattern of activity in a whole group of neurons. We now think of Hebb's cell assembly more generally as a spatiotemporal pattern in the brain which represents an object, an action, or an abstraction such as an idea. Each is like a musical melody and, I calculate, takes up about as much space in the brain as would the head of a pin (just imagine that the pinhead is hexagonal in shape).

Memories are mere spatial patterns frozen in time – that sheet of music waiting for a pianist, or the ruts in a washboarded road, lying in wait for something to come along and interact with them to produce a spatiotemporal pattern in the form of live music or a bouncing car. A Darwinian model of mind suggests that an activated memory can interact with other plans for action, compete for occupation of a work space. A passive memory, like those ruts in the road, can also serve as an aspect of the environment that biases a competition – in short, both the current real-time environment and memories of past environments can bias a competition that shapes up a thought.

So we have a pattern – that musiclike thought in the brain – and we have selective survival biased by a multifaceted environment. How can thoughts be copied to produce dozens of identical pinheads? How can their variants compete for a work space, the same as bluegrass and crabgrass compete for a backyard? How can the loop be closed?

All of the currently active cerebral codes in the brain, whether for objects like apples or for skilled finger movements such as dialing a telephone, are thought to be spatiotemporal patterns. To move a code from one part of the brain to another, it isn't physically sent, as a letter is mailed. Rather, it has to be copied much like a fax machine makes a copy of the pattern on one sheet of paper onto a new sheet of paper at the remote location. The transmission of a neural code involves making a copy of a spatiotemporal pattern, sometimes a distant copy via the fibers of the *corpus callosum* but often a nearby copy, much in the manner that a crystal grows.

The cerebral cortex of the brain, which is where thoughts are most likely to arise, has circuitry for copying spatiotemporal patterns in an immediately adjacent region less than a millimeter away. All primates have this wiring, but it is not known how often they use it. The cerebral cortex is a big sheet – if peeled off and flattened out, it would be about the size of enough pie crust to cover four pies. There are at least 104 standard subdivisions. While some areas of cortex might be committed to full-time specialization, other areas might often support sideways copying and be erasable work space for Darwinian shaping-up processes.

The picture that emerges from theoretical considerations is one of a patchwork quilt, some parts of which enlarge at the expense of their neighbors as one code comes to dominate. As you try to decide whether to pick an apple or an orange from the fruit bowl on the table, the cerebral code for *apple* may be having a copying competition with the one for *orange*. When one code has enough active copies to trip the action circuits, you reach for the apple. But the orange codes aren't entirely banished; they could linger in the background as subconscious thoughts.

When you try to remember someone's name, but without initial success, the candidate codes might continue copying with variations for the next half hour until, suddenly, Jane Smith's name seems to 'pop into your mind.' Our conscious thought may be only the currently dominant pattern in the copying competition, with many other variants competing for dominance (just as the bluegrass competes with the crabgrass for my backyard), one of which will win a moment later when your thoughts seem to shift focus.

The Darwinian process is something of a default mechanism when there is lots of copying going on, and so we might expect a busy brain to use it. Perhaps human thought is more complicated than this, with shortcuts so completely dominating the picture that the Darwinian aspects are minor. Certainly, the language mechanisms in our brain must involve a lot of rule-based shortcuts, judging from how children make relatively sudden transitions from speaking simple sentences to speaking much more complicated ones, during their third year of life. It may be that the Darwinian processes are only the frosting on the cake, that much is routine and rule-bound.

But the frosting isn't just writing poetry or creating scientific theories (such as this one). We often deal with novel situations in creative ways, as when deciding what to fix for dinner tonight. We survey what's already in the refrigerator and on the kitchen shelves. We think about a few alternatives, keeping track of what else we might have to fetch from the grocery store. And we sometimes combine these elements into a stew, or a combination of dishes that we've never had before. All of this can flash though the mind within seconds – and that's probably a Darwinian process at work, as is speculating about what tomorrow might bring.

LEE SMOLIN

What Is Time?

Every schoolchild knows what time is. But, for every schoolchild, there is a moment when they first encounter the paradoxes that lie just behind our everyday understanding of time. I recall when I was a child being struck all of a sudden by the question of whether time could end or whether it must go on forever. It must end, for how can we conceive of the infinity of existence stretching out before us if time is limitless? But if it ends, what happens afterward?

I have been studying the question of what time is for much of my adult life. But I must admit at the beginning that I am no closer to an answer now than I was then. Indeed, even after all this study, I do not think we can answer even the simple question: 'What sort of thing is time?' Perhaps the best thing I can say about time is to explain how the mystery has deepened for me as I have tried to confront it.

Here is another paradox about time which I began to worry about only after growing up. We all know that clocks measure time. But clocks are complex physical systems and hence are subject to imperfection, breakage, and disruptions of electrical power. If I take any two real clocks, synchronize them, and let them run, after some time, they will always disagree about what time it is.

So which of them measures the real time? Indeed, is there a single, absolute time which, although measured imperfectly by any actual clock, is the true time of the world? It seems there must be, otherwise, what do we mean when we say that some particular clock runs slow or fast? On the other hand, what could it mean to say that something like an absolute time exists if it can never be precisely measured?

A belief in an absolute time raises other paradoxes. Would time flow if there were nothing in the universe? If everything stopped, if nothing happened, would time continue?

On the other hand, perhaps there is no single absolute time. In that case, time is only what clocks measure and, as there are many clocks and they all, in the end, disagree, there are many times. Without an absolute time, we can only say that time is defined relative to whichever clock we choose to use.

This seems to be an attractive point of view, because it does not lead us to believe in some absolute flow of time we can't observe. But it leads to a problem, as soon as we know a little science.

One of the things physics describes is motion, and we cannot conceive of motion without time. Thus, the notion of time is basic for physics. Let me take the simplest law of motion, which was invented by Galileo and Descartes, and formalized by Isaac Newton: A body with no forces acting on it moves in a straight line at a constant speed. Let's not worry here about what a straight line is (this is the problem of space, which is perfectly analogous to the problem of time, but which I won't discuss here). To understand what this law is asserting, we need to know what it means to move at a constant speed. This concept involves a notion of time, as one moves at a constant speed when equal distances are covered in equal times.

We may then ask: With respect to which time is the motion to be constant? Is it the time of some particular clock? If so, how do we know which clock? We must certainly choose because, as we observed a moment ago, all real clocks will eventually go out of synchronization with one another. Or is it rather that the law refers to an ideal, absolute time?

Suppose we take the point of view that the law refers to a single, absolute time. This solves the problem of choosing which clock to use, but it raises another problem, for no real, physical clock perfectly measures this imagined, ideal time. How could we truly be sure whether the statement of the law is true, if we have no access to this absolute, ideal time? How do we tell whether some apparent speeding up or slowing down of some body in a particular experiment is due to the failure of the law, or 46 only to the imperfection of the clock we are using?

Newton, when he formulated his laws of motion, chose to solve the problem of which clock by positing the existence of an absolute time. Doing this, he went against the judgments of his contemporaries, such as Descartes and Gottfried Leibniz, who held that time must be only an aspect of the relationships among real things and real processes in the world. Perhaps theirs is the better philosophy, but as Newton knew better than anyone at the time, it was only if one believed in an absolute time that his laws of motion, including the one we have been discussing, make sense. Indeed, Albert Einstein, who overthrew Newton's theory of time, praised Newton's 'courage and judgment' to go against what is clearly the better philosophical argument, and make whatever assumptions he had to make to invent a physics that made sense.

This debate, between time as absolute and preexisting and time as an aspect of the relations of things, can be illustrated in the following way. Imagine that the universe is a stage on which a string quartet or a jazz group is about to perform. The stage and the hall are now empty, but we hear a ticking, as someone has forgotten, after the last rehearsal, to turn off a metronome sitting in a corner of the orchestra pit. The metronome ticking in the empty hall is Newton's imagined absolute time, which proceeds eternally at a fixed rate, prior to and independently of anything actually existing or happening in the universe. The musicians enter, the universe all of a sudden is not 47 empty but is in motion, and they begin to weave their rhythmic art. Now, the time that emerges in their music is not the absolute preexisting time of the metronome; it is a relational time based on the developing real relationships among the musical thoughts and phrases. We know this is so, for the musicians do not listen to the metronome, they listen to one another, and through their musical interchange, they make a time that is unique to their place and moment in the universe.

But, all the while, in its corner the metronome ticks on, unheard by the music makers. For Newton, the time of the musicians, the relational time, is a shadow of the true, absolute time of the metronome. Any heard rhythm, as well as the ticking of any real physical clock, only traces imperfectly the true absolute time. On the other hand, for Leibniz and other philosophers, the metronome is a fantasy that blinds us to what is really happening; the only time is the rhythm the musicians weave together.

The debate between absolute and relational time echoes down the history of physics and philosophy, and confronts us now, at the end of the twentieth century, as we try to understand what notion of space and time is to replace Newton's.

If there is no absolute time, then Newton's laws of motion don't make sense. What must replace them has to be a different kind of law that can make sense if one measures time by any clock. That is, what is required is a
48 democratic rather than an autocratic law, in which any

clock's time, imperfect as it may be, is as good as any other's. Now, Leibniz was never able to invent such a law. But Einstein did, and it is indeed one of the great achievements of his theory of general relativity that a way was found to express the laws of motion so that they make sense whichever clock one uses to embody them with meaning. Paradoxically, this is done by eliminating any reference to time from the basic equations of the theory. The result is that time cannot be spoken about generally or abstractly; we can only describe how the universe changes in time if we first tell the theory exactly which real physical processes are to be used as clocks to measure the passage of time.

So, this much being clear, why then do I say that I do not know what time is? The problem is that general relativity is only half of the revolution of twentieth-century physics, for there is also the quantum theory. And quantum theory, which was originally developed to explain the properties of atoms and molecules, took over completely Newton's notion of an absolute ideal time.

So, in theoretical physics, we have at present not one theory of nature but two theories: relativity and quantum mechanics, and they are based on two different notions of time. The key problem of theoretical physics at the present moment is to combine general relativity and quantum mechanics into one single theory of nature that can finally replace the Newtonian theory overthrown at 49 the beginning of this century. And, indeed, the key obstacle to doing this is that the two theories describe the world in terms of different notions of time.

Unless one wants to go backward and base this unification on the old, Newtonian notion of time, it is clear that the problem is to bring the Leibnizian, relational notion of time into the quantum theory. This is, unfortunately, not so easy. The problem is that quantum mechanics allows many different, and apparently contradictory, situations to exist simultaneously, as long as they exist in a kind of shadow or potential reality. (To explain this, I would have to write another essay at least as long as this one about the quantum theory.) This applies to clocks as well, in the same way that a cat in quantum theory can exist in a state that is at the same time potentially living and potentially dead, a clock can exist in a state in which it is simultaneously running the usual way and running backward. So, if there were a quantum theory of time, it would have to deal not only with freedom to choose different physical clocks to measure time, but with the simultaneous existence, at least potentially, of many different clocks. The first, we have learned from Einstein how to do; the second has, so far, been too much for our imaginations.

So the problem of what time is remains unsolved. But it is worse than this, because relativity theory seems to require other changes in the notion of time. One of them concerns my opening question, whether time can begin or end, or whether it flows eternally. For relativity is a theory in which time can truly start and stop.

One circumstance in which this happens is inside of a black hole. A black hole is the result of the collapse of a massive star, when it has burned all its nuclear fuel and thus ceased to burn as a star. Once it is no longer generating heat, nothing can halt the collapse of a sufficiently massive star under the force of its own gravity. This process feeds on itself, because the smaller the star becomes, the stronger the force by which its parts are mutually attracted to one another. One consequence of this is that a point is reached at which something would have to go faster than light to escape from the surface of the star. Since nothing can travel faster than light, nothing can leave. This is why we call it a black hole, for not even light can escape from it.

However, let us think not of this, but of what happens to the star itself. Once it becomes invisible to us, it takes only a short time for the whole star to be compressed to the point at which it has an infinite density of matter and an infinite gravitational field. The problem is, what happens then? The problem, indeed, is what, in such a circumstance, 'then' might mean. If time is only given meaning by the motion of physical clocks, then we must say that time stops inside of each black hole. Because once the star reaches the state of infinite density and infinite gravitational field, no further change can take place, and no physical process can go on that would give meaning to time. Thus, the theory simply asserts that time stops.

The problem is in fact even worse than this, because general relativity allows for the whole universe to collapse like a black hole, in which case, time stops everywhere. It can also allow for time to begin. This is the way we understand the big bang, the most popular theory, currently, of the origin of the universe.

Perhaps the central problem that those of us who are trying to combine relativity and quantum mechanics think about is what is really happening inside a black hole. If time really stops there, then we must contemplate that all time, everywhere, comes to a stop in the collapse of the universe. On the other hand, if it does not stop, then we must contemplate a whole, limitless world inside each black hole, forever beyond our vision. Moreover, this is not just a theoretical problem, because a black hole is formed each time a massive enough star comes to the end of its life and collapses, and this mystery is occurring, somewhere in the vast universe we can see, perhaps one hundred times a second.

So, what is time? Is it the greatest mystery? No, the greatest mystery must be that each of us is here, for some brief time, and that part of the participation that the universe allows us in its larger existence is to ask such questions. And to pass on, from schoolchild to schoolchild, the joy of wondering, of asking, and of telling each other what we know and what we don't know.

W. DANIEL HILLIS

Special Relativity: Why Can't You Go Faster Than Light?

Y ou've probably heard that nothing can go faster than the speed of light, but have you ever wondered how this rule gets enforced? What happens when you're cruising along in your spaceship and you go faster and faster and faster until you hit the light barrier? Do the dilithium crystals that power your engine suddenly melt down? Do you vanish from the known universe? Do you go backward in time? The correct answer is none of the above. Don't feel bad if you don't know it; no one in the world knew it until Albert Einstein worked it out.

The easiest way to understand Einstein's explanation is to understand the simple equation that you have probably seen before: $e = mc^2$. In order to understand this equation, let's consider a similar equation, one for converting between square inches and square feet. If i is the number of square inches and f is the number of square feet, then we can write the equation: $i = 144f$. The 144 comes from squaring the number of inches per foot ($12^2 = 144$). Another way of writing the same equation would be $i = c^2f$, where c in this case is equal to 12 inches per foot. Depending on what units we use, this equation can be used to convert any measure of area to any other measure of area; just the constant c will be different. For example, the same equation can be used for converting square yards to square meters, where c^2 is 0.9144, the number of yards per meter. The c^2 is just the conversion constant.

The reason why these area equations work is that square feet and square inches are different ways of measuring the same thing, namely area. What Einstein realized, to everyone's surprise, was that energy and mass are also just two different ways of measuring the same thing. It turns out that just a little bit of mass is equal to a whole lot of energy, so in the equation, the conversion constant is very large. For example, if we measure mass in kilograms and energy in joules, the equation can be written like this: $e = 90{,}000{,}000{,}000{,}000{,}000\, m$. This means, for example, that a charged-up battery (which contains about one million joules of energy) weighs about 0.0000000001 grams more than a battery that has been discharged.

If we work with different units, the conversion constant will be different. For instance, if we measure mass in tons, and energy in BTUs, then c will be 93,856,000,000,000,000. (It happens to work out that the conversion constant in a particular set of units is always the speed of light in those units, but that is another story.) If we measure both energy and mass in what physicists call 'the natural units' (in which $c = 1$), we would write the equation: $e = m$, which makes it easier to understand; it just means that energy and mass are the same thing.

It doesn't matter whether the energy is electrical energy, chemical energy, or even atomic energy. It all weighs the same amount per unit of energy. In fact, the equation even works with something physicists called 'kinetic' energy, that is, the energy something has when it is moving. For example, when I throw a baseball, I put energy into the baseball by pushing it with my arm. According to Einstein's equation, the baseball actually gets heavier when I throw it. (A physicist might get picky here and distinguish between something getting heavier and something gaining mass, but I'm not going to try. The point is that the ball becomes harder to throw.) The faster I throw the baseball, the heavier it gets. Using Einstein's equation, $e = mc^2$, I calculate that if I could throw a baseball one hundred miles an hour (which I can't, but a good pitcher can), then the baseball actually gets heavier by 0.000000000002 grams – which is not much.

Now, let's go back to your starship. Let's assume that your engines are powered by tapping into some external energy source, so you don't have to worry about carrying fuel. As you get going faster and faster in your starship, you are putting more and more energy into the ship by speeding it up, so the ship keeps getting heavier. (Again, I should really be saying 'massier' not 'heavier' since there is no gravity in space.) By the time you reach 90 percent of the speed of light, the ship has so much energy in it that it actually has about twice the mass as the ship has at rest. It gets harder and harder to propel with the engines, because it's so heavy. As you get closer to the speed of light, you begin to get diminishing returns – the more energy the ship has, the heavier it gets, so the more energy that must be put into it to speed it up just a little bit, the heavier it gets, and so on.

The effect is even worse than you might think because of what is going on inside the ship. After all, everything inside the ship, including you, is speeding up, getting more and more energy, and getting heavier and heavier. In fact, you and all the machines on the ship are getting pretty sluggish. Your watch, for instance, which used to weigh about half an ounce, now weighs about forty tons. And the spring inside your watch really hasn't gotten any stronger, so the watch has slowed way down so that it only ticks once an hour. Not only has your watch slowed down, but the biological clock inside your head has also slowed down. You don't notice this because your neurons are getting heavier, and your thoughts are slowed down by exactly the same amount as the watch. As far as you are concerned, your watch is just ticking along at the same rate as before. (Physicists call this 'relativistic time contraction.') The other thing that is slowed down is all of the machinery that is powering your engines (the dilithium crystals are getting heavier and slower, too). So your ship is getting heavier, your engines are getting sluggish, and the closer you get to the speed of light, the worse it gets. It just gets harder and harder and harder, and no matter how hard you try, you just can't quite get over the light barrier. And that's why you can't go faster than the speed of light.

PHOENIX 60P PAPERBACKS

HISTORY/BIOGRAPY/TRAVEL
The Empire of Rome A.D. 98–190 *Edward Gibbon*
The Prince *Machiavelli*
The Alan Clark Diaries: Thatcher's Fall *Alan Clark*
Churchill: Embattled Hero *Andrew Roberts*
The French Revolution *E.J. Hobsbawm*
Voyage Around the Horn *Joshua Slocum*
The Great Fire of London *Samuel Pepys*
Utopia *Thomas More*
The Holocaust *Paul Johnson*
Tolstoy and History *Isaiah Berlin*

SCIENCE AND PHILOSOPHY
A Guide to Happiness *Epicurus*
Natural Selection *Charles Darwin*
Science, Mind & Cosmos *John Brockman, ed.*
Zarathustra *Friedrich Nietzsche*
God's Utility Function *Richard Dawkins*
Human Origins *Richard Leakey*
Sophie's World: The Greek Philosophers *Jostein Gaarder*
The Rights of Woman *Mary Wollstonecraft*
The Communist Manifesto *Karl Marx & Friedrich Engels*
Birds of Heaven *Ben Okri*

FICTION
Riot at Misri Mandi *Vikram Seth*
The Time Machine *H. G. Wells*
Love in the Night *F. Scott Fitzgerald*

The Murders in the Rue Morgue *Edgar Allan Poe*
The Necklace *Guy de Maupassant*
You Touched Me *D. H. Lawrence*
The Mabinogion *Anon*
Mowgli's Brothers *Rudyard Kipling*
Shancarrig *Maeve Binchy*
A Voyage to Lilliput *Jonathan Swift*

POETRY
Songs of Innocence and Experience *William Blake*
The Eve of Saint Agnes *John Keats*
High Waving Heather *The Brontes*
Sailing to Byzantium *W. B. Yeats*
I Sing the Body Electric *Walt Whitman*
The Ancient Mariner *Samuel Taylor Coleridge*
Intimations of Immortality *William Wordsworth*
Palgrave's Golden Treasury of Love Poems *Francis Palgrave*
Goblin Market *Christina Rossetti*
Fern Hill *Dylan Thomas*

LITERATURE OF PASSION
Don Juan *Lord Byron*
From Bed to Bed *Catullus*
Satyricon *Petronius*
Love Poems *John Donne*
Portrait of a Marriage *Nigel Nicolson*
The Ballad of Reading Gaol *Oscar Wilde*
Love Sonnets *William Shakespeare*
Fanny Hill *John Cleland*
The Sexual Labyrinth (for women) *Alina Reyes*
Close Encounters (for men) *Alina Reyes*